A Walk In The Woods

Produced by Daniel Weiss Associates, Inc.
33 West 17 Street, New York, NY 10011

Published by Silver Press, a division of
Silver Burdett Press, Inc., Simon & Schuster, Inc.
Prentice Hall Bldg., Englewood Cliffs, NJ 07632
For information address: Silver Press.

Printed in the United States of America
10 9 8 7 6 5 4 3 2 1

Library of Congress Cataloging-in-Publication Data

Arnold, Caroline.
A walk in the woods/written by Caroline Arnold; illustrated
by Freya Tanz.
p. cm.—(First facts)
Summary: Describes some of the plants and animals that live in
the forest and how the forest changes with each new season.
1. Forest ecology—juvenile literature. 2. Forest fauna—juvenile
literature. 3. Forest flora—juvenile literature. [1. Forest ecology.
2. Forest animals. 3. Forest plants. 4. Ecology.] I. Tanz, Freya, ill.
II. Title. III. Series: First facts.
(Englewood Cliffs, N.J.)

QH541.5.F6A75 1990   90-8335
574.5.2642—dc20   CIP
                                                                             AC
ISBN 0-671-68665-8   ISBN 0-671-68661-5 (lib. bdg.)

First Facts ™

A Walk In The Woods

Written by Caroline Arnold
Illustrated by Freya Tanz

Silver Press

See the sun peeking over the treetops.
Feel the damp earth under your feet.
Listen to the birds calling overhead.
Where are we?

We're in a forest!
Let's take a walk and see what we can find.

Wild blackberry and blueberry bushes
need a lot of sunlight.
They grow best at the edge of the forest.

Chirrup, chirrup, call the robins.
They love to eat berries.
Let's follow them into the forest.
We'll go where the trees grow tall.

Trees, trees all around.
Some are tall. Some are wide.
All kinds of trees grow in the forest.
Feel the smooth bark of the birch tree.
Rub the shaggy hickory bark.
Every tree has its own kind of bark.

And each has different shaped leaves.
Maple leaves have five large points.
Oak leaves have round lobes.
And elm leaves have jagged edges.

Oak

Maple

Elm

Pine trees do not have leaves.
They have needles instead.
Ouch! Be careful when you touch them.
Pine needles are prickly.

Scotch Pine

White Pine

Spruce

Balsam Fir

Pick up a pine cone.
Can you see its tiny seeds?
When the seeds fall to the ground,
some may grow into new pine trees.
All trees have seeds.

Seeds are eaten by many forest animals.
Watch squirrels bury acorns in the ground.
If they don't come back to eat them,
the acorns may become new oak trees.

Seeds sprout roots as they
begin to grow.
The roots grow deep into the ground.
They take in water and
food from the soil.

How can you tell the age of a tree?
Look at this wide tree stump.
A new pair of rings is added every year.
The light ring grows in spring.
The dark one grows in summer.
Try to count them. About how old was this tree?

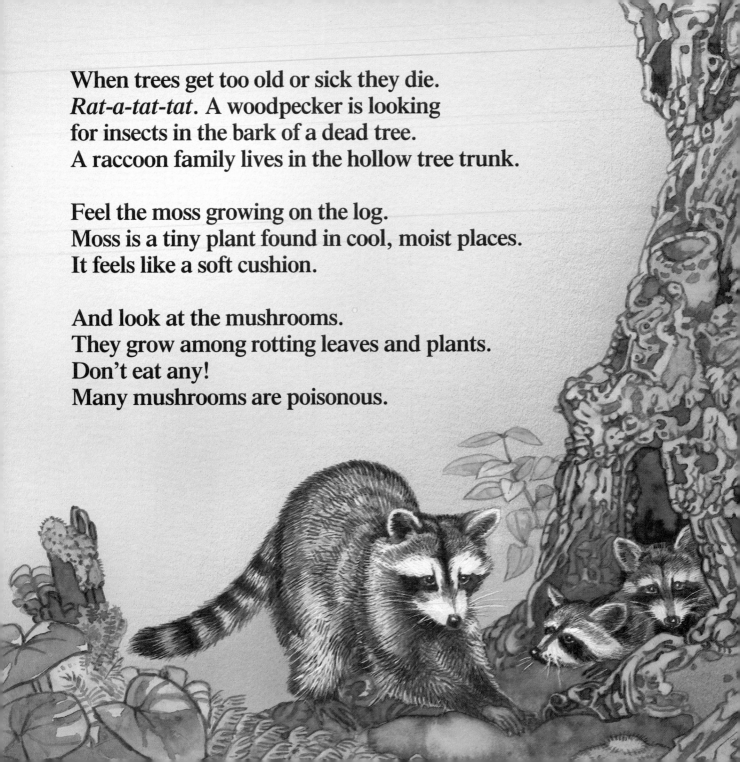

When trees get too old or sick they die.
Rat-a-tat-tat. A woodpecker is looking
for insects in the bark of a dead tree.
A raccoon family lives in the hollow tree trunk.

Feel the moss growing on the log.
Moss is a tiny plant found in cool, moist places.
It feels like a soft cushion.

And look at the mushrooms.
They grow among rotting leaves and plants.
Don't eat any!
Many mushrooms are poisonous.

Crash! Bang! What's that noise?
Beavers are busy cutting down trees.
They are building a dam.
The dam makes a deep pond where the beavers
can build their lodge.

The pond is home to fish, plants, and insects.
These become food for other pond animals.

In the forest meadow, there are no trees.
But there are many bright flowers and
colorful butterflies.

Other creatures blend in
with the colors around them.
This is called camouflage.
Camouflage helps creatures stay hidden.

Can you find the garter snake?
Watch it slither through the tall grass.
It is going toward the pond to find
frogs, worms, and fish to eat.

Look at the mother rabbit and her bunnies.
They are looking for plants to eat.
They must watch out for the hungry fox.

The mother pheasant is also on the lookout.
Her nest is well hidden in the tall grass.
Soon her eggs will hatch.
The baby pheasants will grow by eating
seeds and insects.
All the baby animals are bigger by summer's end.

The green leaves are turning red and yellow.
It is autumn.
Wild nuts and fruits ripen
and fall to the ground.
Animals feast on these before food becomes scarce.

Many birds fly south to warmer weather.
Honk, *honk*, cry the geese as they fly by.

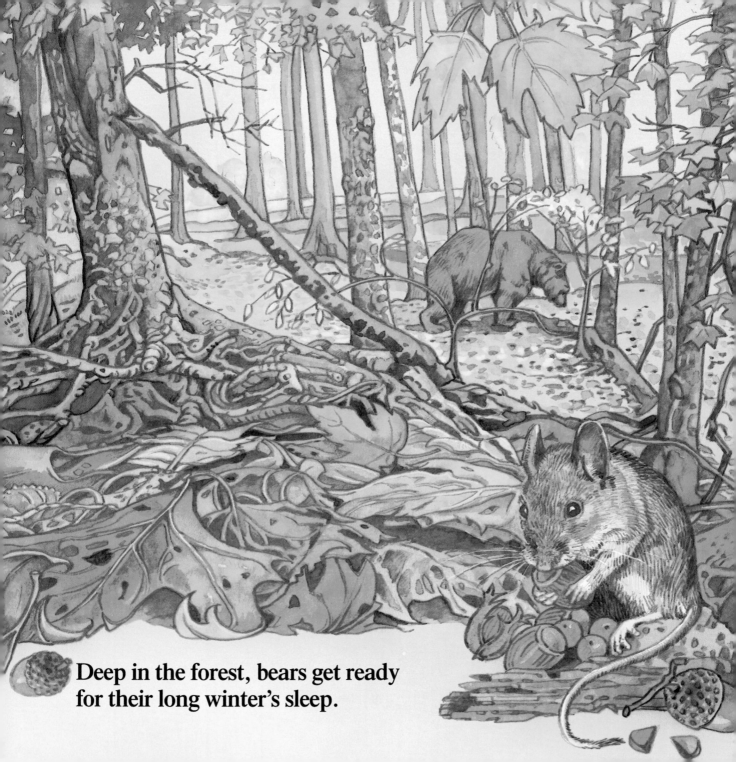

Deep in the forest, bears get ready
for their long winter's sleep.

In winter, snow covers the ground.
The pond freezes over.
Many of the trees are bare.
Only the pine trees stay green.
Trees that stay green all year
are called evergreens.

Just a few animals come out in winter.
Can you follow these footprints across the snow?

When spring comes again,
tree buds swell and new leaves grow.
Bears come out of their dens.
Birds return to build their nests.
Many baby animals are born.
It is the start of a new forest year.

We cannot live without forests.
Trees give us wood to build
houses and furniture.

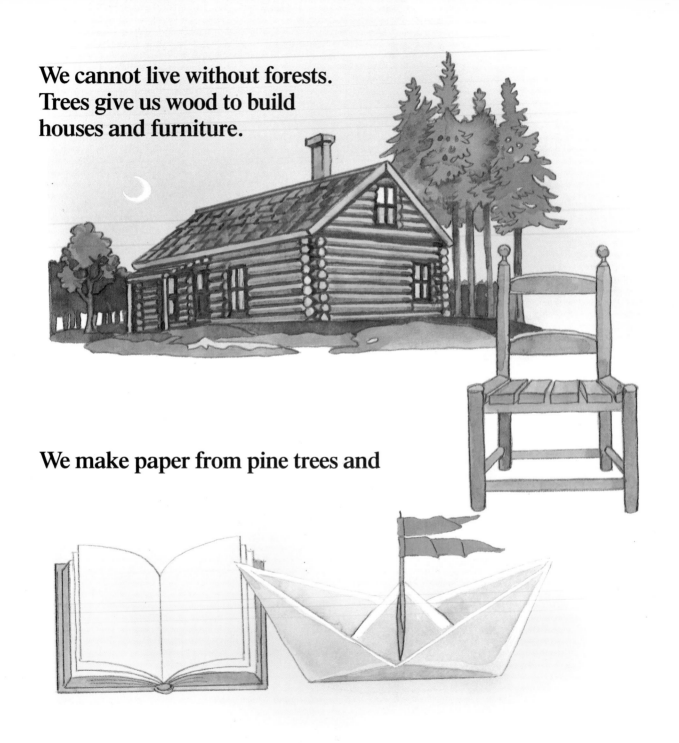

We make paper from pine trees and

syrup from maple trees.

Trees also give us shade.

You can find forests all over the world.

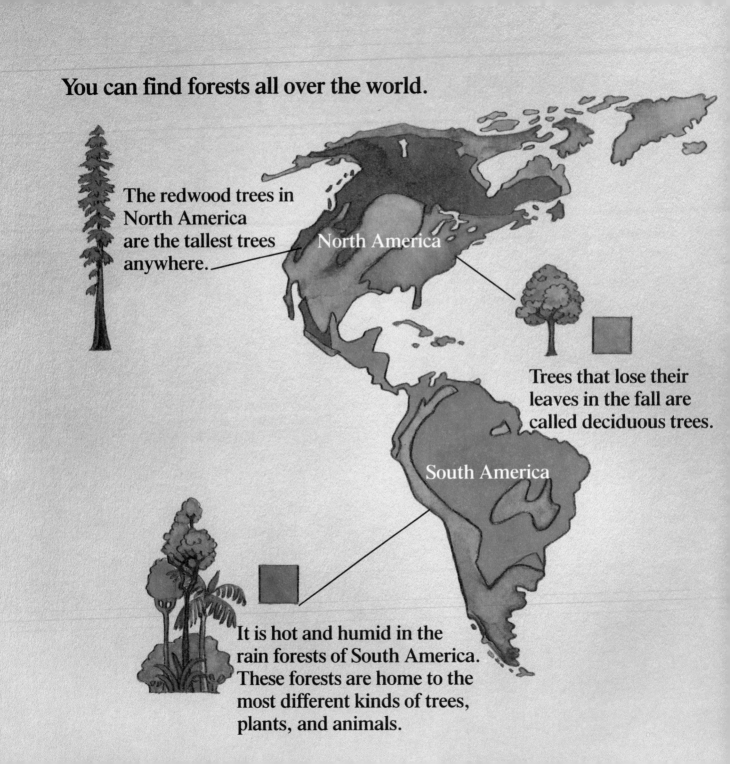

The redwood trees in North America are the tallest trees anywhere.

North America

Trees that lose their leaves in the fall are called deciduous trees.

South America

It is hot and humid in the rain forests of South America. These forests are home to the most different kinds of trees, plants, and animals.

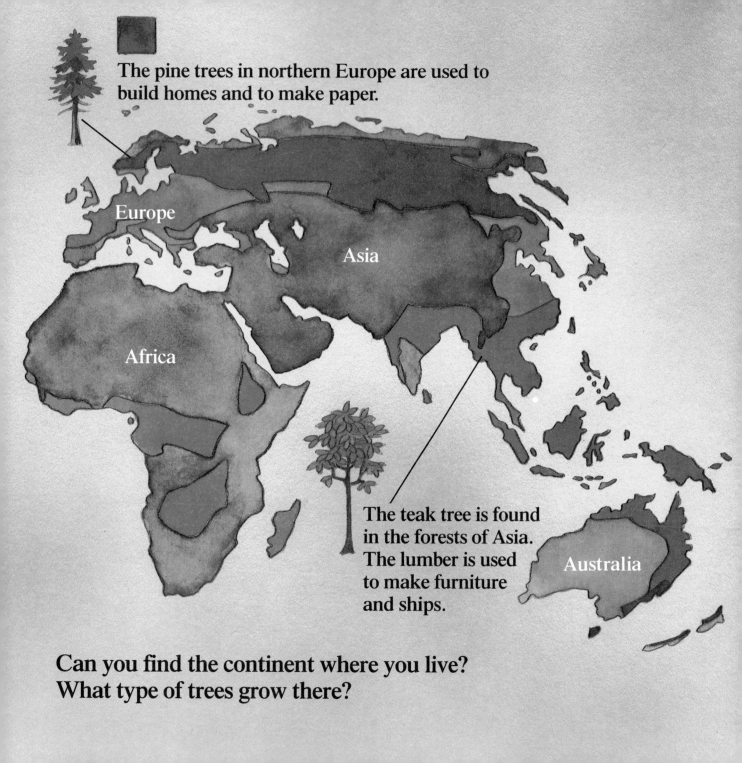

The pine trees in northern Europe are used to build homes and to make paper.

Europe

Asia

Africa

The teak tree is found in the forests of Asia. The lumber is used to make furniture and ships.

Australia

Can you find the continent where you live?
What type of trees grow there?

What did you see on your walk in the woods?
Here are some clues to help you remember.
Can you name each one?